海王星

风暴最多的星球

NEPTUNE

The Stormiest Planet

（英国）埃伦·劳伦斯／著　　张 骁／译

江苏凤凰美术出版社

著作权合同登记图字：10-2022-144

图书在版编目（CIP）数据

海王星：风暴最多的星球 /（英）埃伦·劳伦斯著；
张骁译 . -- 南京：江苏凤凰美术出版社，2025. 1.
（环游太空）. -- ISBN 978-7-5741-2027-3

Ⅰ . P185.6-49

中国国家版本馆 CIP 数据核字第 2024RM0420 号

策　　　　划	朱　婧	
责 任 编 辑	高　静	奚　鑫
责 任 校 对	王　璇	
责任设计编辑	樊旭颖	
责 任 监 印	生　嫄	
英 文 朗 读	C.A.Scully	
项 目 协 助	邵楚楚	乔一文雯

丛 　书 　名	环游太空
书 　　　名	海王星：风暴最多的星球
著 　　　者	（英国）埃伦·劳伦斯
译 　　　者	张　骁
出 版 发 行	江苏凤凰美术出版社（南京市湖南路1号 邮编：210009）
印 　　　刷	南京新世纪联盟印务有限公司
开 　　　本	710 mm×1000 mm　1/16
总 　印 　张	18
版 　　　次	2025 年 1 月第 1 版
印 　　　次	2025 年 1 月第 1 次印刷
标 准 书 号	ISBN 978-7-5741-2027-3
总 　定 　价	198.00 元（全 12 册）

营销部电话：025-68155675　营销部地址：南京市湖南路1号
江苏凤凰美术出版社图书凡印装错误可向承印厂调换

目录 Contents

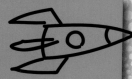

书中加粗的词语见词汇表解释。

Words shown in **bold** in the text are explained in the glossary.

欢迎来到海王星
Welcome to Neptune

想象一下，你正在飞往一个离地球好几十亿千米的世界。

Imagine flying to a world that is billions of kilometers from Earth.

随着飞船越来越近，你看到了一个蓝色的星球，上面覆盖着厚厚的云层。

As your spacecraft gets close, you see a blue world covered with clouds.

云层中的狂怒的风暴，能够随时摧毁你的飞船。

Inside the clouds are strong winds and fierce storms that could destroy your spacecraft.

你成功地飞过了云层，但是却根本没有地方着陆。

You manage to fly through the clouds, but there is nowhere to land.

因为这个遥远的巨大球体其实是由极寒的气体和液体组成的。

That's because this faraway world is a huge ball of icy **gases** and liquids.

欢迎来到行星海王星！

Welcome to the **planet** Neptune!

人类从未到达过海王星，但是有一个飞行器做到了。1989年，一个叫作"旅行者2号"的太空探测器飞越了这颗行星。

No humans have ever visited Neptune, but a spacecraft has. In 1989, a space **probe** called *Voyager 2* flew past the planet.

这张海王星的照片就是由"旅行者2号"拍摄的。
This photograph of Neptune was taken by *Voyager 2.*

太阳系 The Solar System

海王星以超过19 000千米每小时的速度在太空移动。

Neptune is moving through space at over 19,000 kilometers per hour.

它围绕着太阳做一个巨大的圆周运动。

It is moving in a huge circle around the Sun.

海王星是围绕太阳公转的八大行星之一。

Neptune is one of eight planets circling the Sun.

八大行星分别是水星、金星、我们的母星地球、火星、木星、土星、天王星和海王星。

The planets are called Mercury, Venus, our home planet Earth, Mars, Jupiter, Saturn, Uranus, and Neptune.

冰冻的彗星和被称为"小行星"的大型岩石也围绕着太阳公转。

Icy **comets** and large rocks, called **asteroids**, are also moving around the Sun.

太阳、行星和其他天体共同组成了"太阳系"。

Together, the Sun, the planets, and other space objects are called the **solar system**.

太阳系中的大多数小行星都集中在被称为"小行星带"的环状带中。

Most of the asteroids in the solar system are in a ring called the asteroid belt.

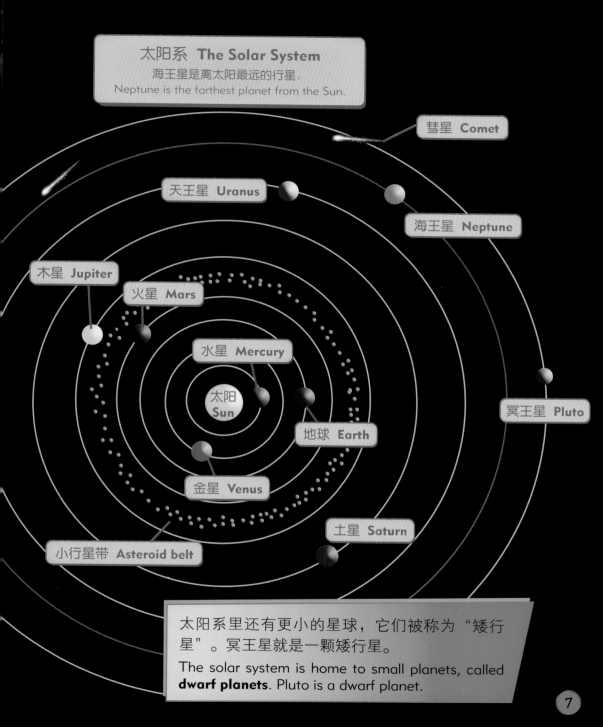

太阳系 The Solar System

海王星是离太阳最远的行星。
Neptune is the farthest planet from the Sun.

彗星 Comet

天王星 Uranus

海王星 Neptune

木星 Jupiter

火星 Mars

水星 Mercury

太阳 Sun

地球 Earth

金星 Venus

冥王星 Pluto

小行星带 Asteroid belt

土星 Saturn

太阳系里还有更小的星球，它们被称为"矮行星"。冥王星就是一颗矮行星。

The solar system is home to small planets, called **dwarf planets**. Pluto is a dwarf planet.

海王星的奇幻之旅
Neptune's Amazing Journey

行星围绕太阳公转一圈所需的时间被称为"一年"。

The time it takes a planet to **orbit**, or circle, the Sun once is called its year.

地球绕太阳公转一圈需要略多于365天，所以地球上的一年有365天。

Earth takes just over 365 days to orbit the Sun, so a year on Earth lasts 365 days.

海王星比地球离太阳更远，所以它绕太阳公转的路程更长。

Neptune is farther from the Sun than Earth, so it must make a much longer journey.

海王星绕太阳公转一周大约需要165个地球年。

It takes Neptune almost 165 Earth years to orbit the Sun.

这意味着，海王星上的一年相当于地球上的165年！

This means that one year on Neptune lasts for 165 Earth years!

当行星围绕太阳公转时，它也像陀螺一样自转着。

As a planet orbits the Sun, it also spins, or **rotates**, like a top.

海王星 Neptune

地球绕太阳公转一圈的路程约为9.4亿千米，而海王星绕太阳公转一圈的距离超过280亿千米！

To orbit the Sun once, Earth makes a journey of about 940 million kilometers. Neptune must make a journey of over 28 billion kilometers!

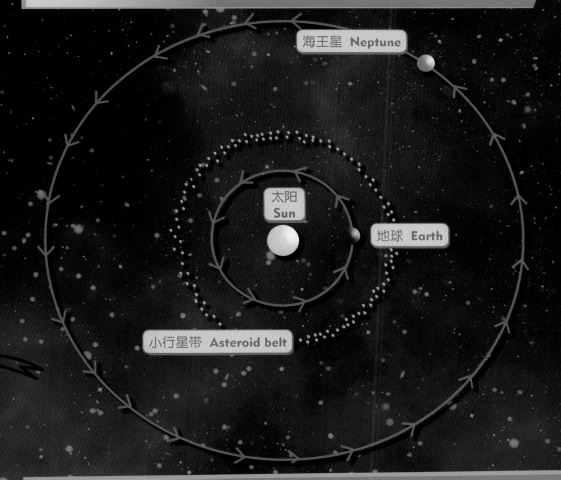

海王星 Neptune

太阳
Sun

地球 Earth

小行星带 Asteroid belt

在这张图片上，我们去掉了太阳系中除了地球和海王星以外的其他行星。现在可以看出海王星绕太阳公转的超长旅程了吧！

In this picture, we've taken away all the planets in the solar system except for Earth and Neptune. Now it's easy to see Neptune's super-long journey around the Sun!

近距离观察海王星
A Closer Look at Neptune

海王星是太阳系中第四大行星。

Neptune is the fourth-largest planet in the solar system.

海王星的直径大约是地球的4倍。

It is about four times wider than Earth.

不同于地球是一颗固态行星，海王星没有固体表面。

Unlike Earth, which is a rocky planet, Neptune doesn't have a solid surface.

这颗行星被一层厚厚的气体和云层包围，它们被称为"大气层"。

The planet is surrounded by a thick layer of gases and clouds called an **atmosphere**.

在大气层之下，这颗行星是一颗由极寒液体组成的巨大球体。

Beneath its atmosphere, the planet is a huge ball of icy liquids.

太阳系的四大行星 The Solar System's Four Largest Planets

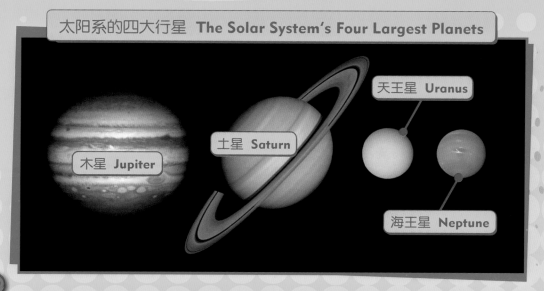

木星 Jupiter

土星 Saturn

天王星 Uranus

海王星 Neptune

10

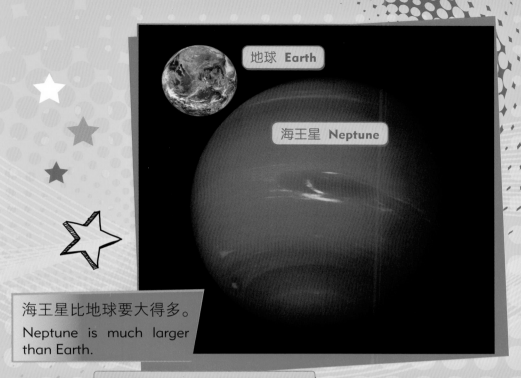

地球 Earth

海王星 Neptune

海王星比地球要大得多。

Neptune is much larger than Earth.

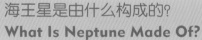

海王星是由什么构成的?
What Is Neptune Made Of?

科学家认为，海王星的中心可能是一个由岩石和冰构成的球体。不过，这个猜想尚未得到证明。

Scientists think there may be a ball of rock and ice in the center of Neptune. No one knows for sure, though.

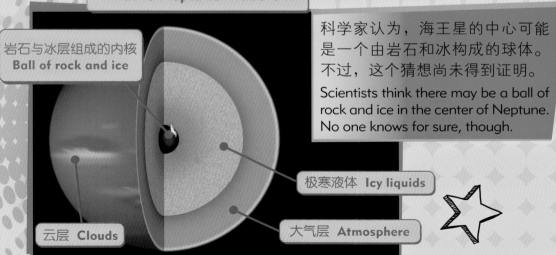

岩石与冰层组成的内核
Ball of rock and ice

极寒液体 Icy liquids

云层 Clouds

大气层 Atmosphere

充满风暴的海王星
Stormy Neptune

显然，海王星是太阳系中风暴最多的行星。

在它的大气层内部有巨型风暴和速度超快的飓风。

这些风的速度可达每小时2 400千米。

这些风暴比地球上最强的飓风还要强大10倍！

当"旅行者2号"探测器到访海王星的时候，它拍下了一张超大飓风的照片。

这个飓风被命名为"大黑斑"。

Neptune is easily the stormiest planet in the solar system.

Inside its atmosphere there are giant storms with super-fast winds.

The winds can blow at 2,400 kilometers per hour.

These storms are 10 times stronger than the most powerful hurricanes on Earth!

When the *Voyager* 2 probe visited Neptune, it took photos of a gigantic hurricane.

The hurricane was called the Great Dark Spot.

云层 Clouds

这张特写照片记录下了云层在海王星上空飞驰而过的样子。

This is a close-up photo of clouds zooming around Neptune.

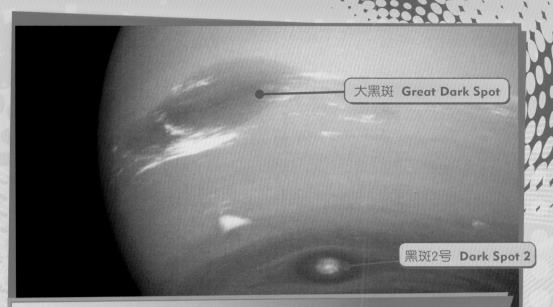

大黑斑 **Great Dark Spot**

黑斑2号 **Dark Spot 2**

这张照片是由"旅行者2号"拍摄的，它显示了大黑斑风暴和一个较小的风暴，被称为"黑斑2号"。

This photo was taken by *Voyager 2*. It shows the Great Dark Spot storm and a smaller storm called Dark Spot 2.

大黑斑真的很大，甚至能装下整个地球！

The Great Dark Spot was so big, planet Earth could fit inside it!

大黑斑 **Great Dark Spot**

海王星的卫星家族
Neptune's Family of Moons

海王星有一个庞大的卫星家族，它们由极寒的岩石组成，围绕着海王星旋转。

Neptune has a large family of icy, rocky **moons** orbiting around it.

我们的家园——地球，只有一颗卫星。

Our home planet, Earth, has just one moon.

海王星至少有16颗卫星，而且科学家们认为也许还会有更多！

Neptune has at least 16 moons, and scientists think there may be more!

海王星最大的卫星——崔顿（海卫一）是太阳系中最为寒冷的地方之一。

Neptune's largest moon, Triton, is one of the coldest places in the solar system.

这颗卫星有着一层固态外壳，是由冰组成的。

This moon has a solid outer crust made of ice.

在外壳之下、卫星内部，是温度极低的液体。

Beneath the crust and inside the moon, there are icy liquids.

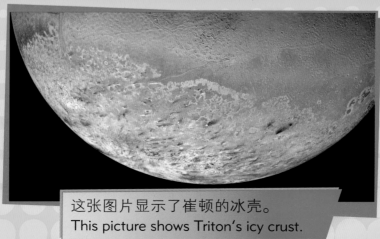

这张图片显示了崔顿的冰壳。
This picture shows Triton's icy crust.

这张图片显示了你可能会在崔顿上看到的场景。这颗非常寒冷的卫星表面有冰火山，它会从崔顿的内部喷射出极寒的液体，直冲天空。

This picture shows what it might look like on Triton. The super-cold moon has ice **volcanoes** on its surface. The volcanoes blast icy liquid from inside Triton up into the sky.

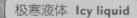 极寒液体 Icy liquid

海王星 Neptune

冰火山 Ice volcano

从崔顿的冰火山中喷出的液体会在卫星地表之上的高空冷却，然后以雪的形态重新降落回来！

The liquid from Triton's ice volcanoes freezes high above the moon's surface. Then it falls back onto Triton as snow!

15

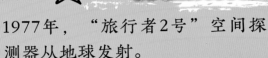

探测海王星的任务
A Mission to Neptune

1977年，"旅行者2号"空间探测器从地球发射。

In 1977, the *Voyager 2* space probe blasted off from Earth.

它要去太阳系中最远的行星执行探测任务。

It was on a mission to visit the solar system's most faraway planets.

"旅行者2号"首先造访了木星，然后是土星和天王星。

First *Voyager 2* visited Jupiter, and then Saturn and Uranus.

1989年，在宇宙中飞行了12年以后，它终于到达了海王星。

In 1989, after 12 years in space, it reached Neptune.

"旅行者2号"乘坐火箭从地球发射，然后它与火箭分离，独自飞往太空。

Voyager 2 blasted off from Earth aboard a rocket. Then it separated from the rocket and flew into space.

火箭 Rocket

为了到达海王星，"旅行者2号"走了将近70亿千米。

To reach Neptune, *Voyager 2* traveled nearly 7 billion kilometers.

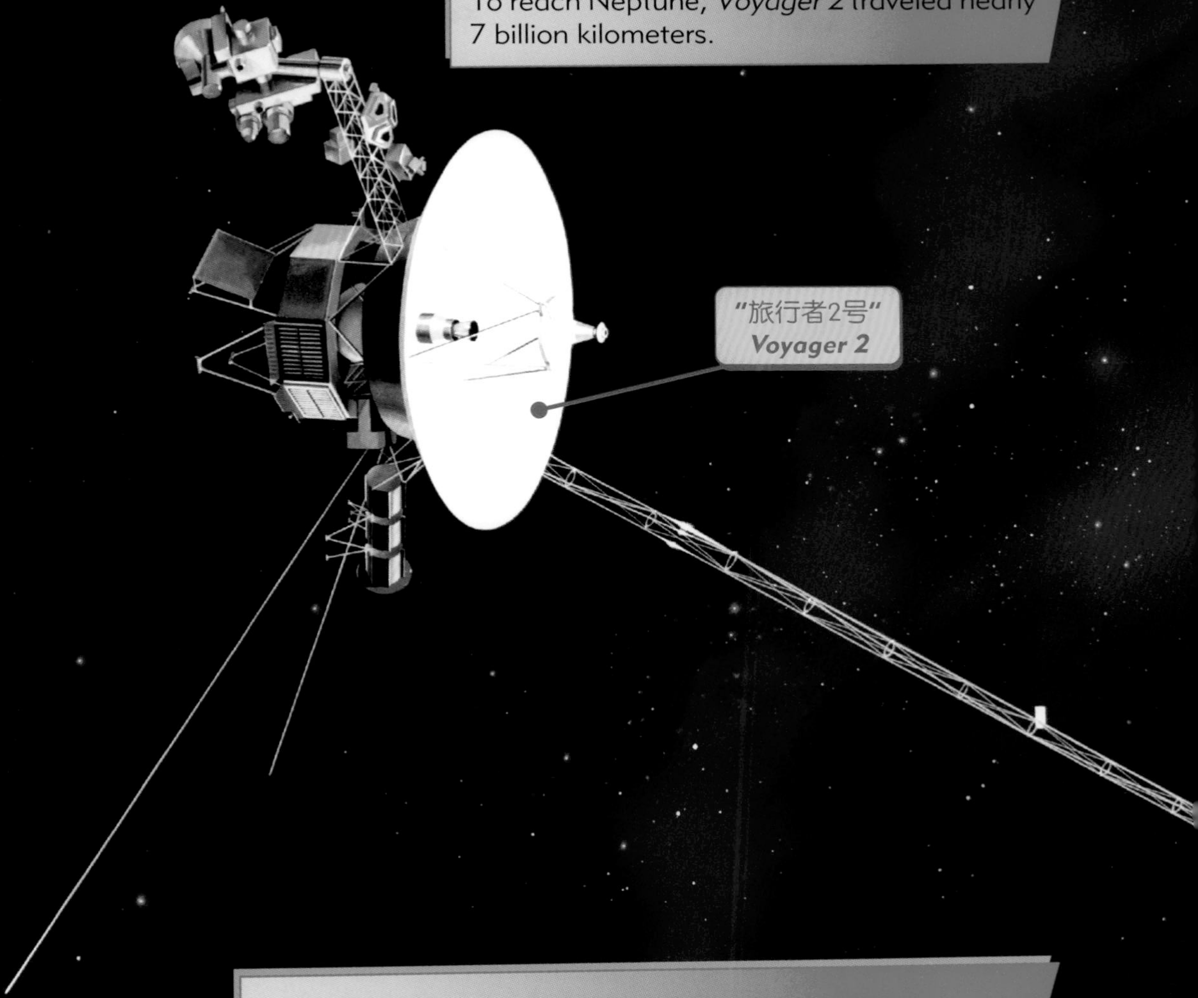

"旅行者2号"
Voyager 2

这张图片展示了探测器在宇宙中飞行时的样子。

This picture shows what the probe might look like flying through space.

激动人心的大发现
Exciting Discoveries

多年以来，科学家认为他们通过望远镜看到了海王星周围的行星环。

For many years, scientists thought they saw rings around Neptune through their telescopes.

当"旅行者2号"到达海王星的时候，这一猜想终于得到了证实。

When *Voyager 2* reached Neptune, it proved they were right.

该探测器拍下了围绕着海王星的行星环照片，这些行星环由小块岩石和尘埃构成。

The probe took photos of rocky, dusty rings circling the planet.

它还发现了6个围绕海王星运行的新卫星。

It also discovered 6 new moons orbiting Neptune.

现在，"旅行者2号"正在飞离海王星。

Voyager 2 is now flying far beyond Neptune.

截止到目前，它是唯一一个到达过海王星的空间探测器。

As of now, it is the only space probe to have visited Neptune.

行星环 Rings

"旅行者2号"拍摄了这张海王星行星环的照片。
Voyager 2 took this photo of Neptune's rings.

"旅行者2号"拍摄了几百张海王星的照片，然后把它们陆陆续续传回地球。这就是当它飞离这颗行星时拍摄的最后几张照片之一。

Voyager 2 took hundreds of photos of Neptune. Then it beamed them back to Earth. This is one of the last photos it took as it flew away from the planet.

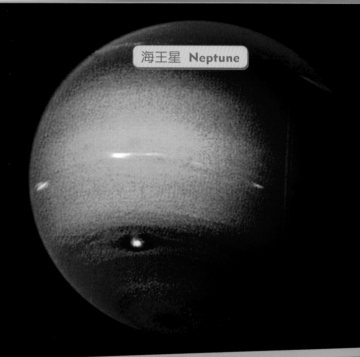

海王星 Neptune

科学家们认为，海王星上会下钻石雨！2017年，科学家们在地球上进行了一项实验。实验表明，海王星大气层中的甲烷气体会变成固体钻石。这种现象同样也会出现在天王星上。

Scientists think that raindrops made of diamonds fall on Neptune! In 2017, scientists on Earth carried out an experiment. It showed them that methane gas in Neptune's atmosphere turns into solid diamonds. The same thing happens on Uranus.

19

有趣的海王星知识
Neptune Fact File

以下是一些有趣的海王星知识：海王星是太阳系中的第八大行星。

Here are some key facts about Neptune, the eighth planet from the Sun.

海王星的发现
Discovery of Neptune

海王星是在1846年被三位科学家发现的，他们分别是奥本·勒维耶、约翰·伽勒和约翰·柯西·亚当斯。

Neptune was discovered in 1846 by three scientists, Urbain Le Verrier, Johann Galle, and John Couch Adams.

海王星是如何得名的
How Neptune got its name

这颗行星是以古罗马神话中的海神命名。

The planet is named after the Roman god of the sea.

行星的大小
Planet sizes

这张图显示了太阳系八大行星的大小对比。

This picture shows the sizes of the solar system's planets compared to each other.

太阳 Sun

水星 Mercury

地球 Earth

金星 Venus

火星 Mars

木星 Jupiter

土星 Saturn

天王星 Uranus

海王星 Neptune

海王星的大小
Neptune's size

海王星的直径约49 244千米
About 49,244 km across

海王星自转一圈需要多长时间
How long it takes for Neptune to rotate once

大约16个地球时
About 16 Earth hours

海王星与太阳的距离
Neptune's distance from the Sun

海王星与太阳的最短距离是4 459 753 056千米。
海王星与太阳的最远距离是4 537 039 826千米。

The closest Neptune gets to the Sun is 4,459,753,056 km.
The farthest Neptune gets from the Sun is 4,537,039,826 km.

海王星围绕太阳公转的平均速度
Average speed at which Neptune orbits the Sun

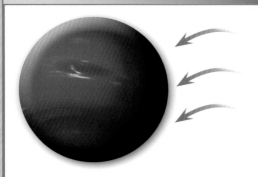

每小时19 566千米
19,566 km/h

海王星绕太阳的轨道长度
Length of Neptune's orbit around the Sun

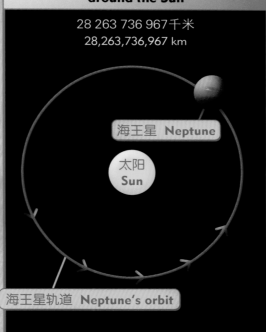

28 263 736 967千米
28,263,736,967 km

海王星 Neptune

太阳
Sun

海王星轨道 Neptune's orbit

海王星上的一年
Length of a year on Neptune

大约60 190个地球天
（大约165个地球年）
About 60,190 Earth days
(about 165 Earth years)

海王星的卫星
Neptune's moons

已知海王星至少有16颗卫星。
可能还有更多的有待发现。

Neptune has at least 16 known moons.
There are possibly more to be discovered.

海王星上的温度
Temperature on Neptune

零下214摄氏度
-214 °C

动动手吧：
熔化的蜡笔海王星图
Get Crafty : Melted Wax Crayon Neptune

使用熔化的蜡笔，在纸上绘制海王星及其充满风暴的云层。

你需要：
- 蜡纸
- 剪刀
- 蓝色和紫色的蜡笔
- 奶酪刨
- 熨斗
  （在成年人的帮助下使用）

1. 从蜡纸上剪下两个餐盘大小的圆。

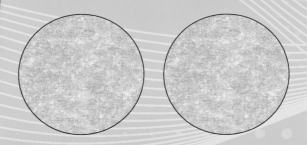

2. 使用奶酪刨将蜡笔屑刨在其中一张圆形蜡纸上。须在成年人的监督下进行操作，务必注意手指的安全！

3. 将另一张蜡纸盖上去，做成蜡笔"三明治"。

4. 请爸爸妈妈帮忙熨烫蜡笔"三明治"，直至蜡笔屑熔化并混合在一起。

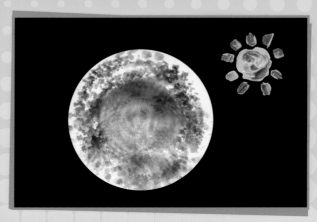

5. 将熔化的蜡笔海王星图挂在窗前，沐浴在阳光下。或者用它制作一幅星空图。

词汇表 Glossary

小行星 | asteroid

围绕太阳公转的大块岩石，有些小得像辆汽车，有些大得像座山。

大气层 | atmosphere

行星、卫星或恒星周围的一层气体。

彗星 | comet

由冰、岩石和尘埃组成的天体，围绕太阳公转。

矮行星 | dwarf planet

围绕太阳运行的圆形或近圆形天体，比八大行星小得多。

气体 | gas

无固定形状或大小的物质，如氧气或氦气。

卫星 | moon

围绕行星运行的天体。通常由岩石或岩石和冰构成。直径从几千米到几百千米不等。地球有一个卫星，名为"月球"。

公转 | orbit

围绕另一个天体运行。

行星 | planet

围绕太阳公转的大型天体：一些行星，如地球，主要是由岩石组成的；其他的行星，如木星，主要是由气体和液体组成的。

探测器 | probe

不载人太空飞船。通常被送往行星或其他天体，用于拍摄照片并收集信息，由地球上的科学家操作控制。

自转 | rotate

物体自行旋转的运动。

太阳系 | solar system

太阳和围绕太阳公转的所有天体，如行星及其卫星、小行星和彗星。

火山 | volcano

地下岩浆喷出地表形成的山丘，部分火山会有高温的液态岩石和气体从开口处喷发；存在于行星或其他天体上。